FAR OUT FACTORIES

MALCOLM E. WEISS

FAR OUT FACTORIES

MANUFACTURING IN SPACE

ILLUSTRATED WITH PHOTOGRAPHS

LODESTAR BOOKS
E. P. DUTTON NEW YORK

LIBRARY OF CONGRESS CATALOGING IN PUBLICATION DATA

Weiss, Malcolm E.
Far out factories.
"Lodestar books."
Includes index.

Summary: Describes the building, operation, and uses of space factories, a concept made possible by the successful flights of the space shuttle.
1. Space stations—Industrial applications—Juvenile literature. [1. Space stations—Industrial applications]
I. Title.
TL797.W43 1984 670.42'0919 83-9030
ISBN 0-525-67143-9

Published in the United States by E. P. Dutton, Inc.,
2 Park Avenue, New York, N.Y. 10016

Published simultaneously in Canada by
Fitzhenry & Whiteside Limited, Toronto

Editor: Virginia Buckley

Printed in the U.S.A. COBE First Edition

10 9 8 7 6 5 4 3 2 1

for Kathie and Kenyon
with much love

CONTENTS

CONVERSION TABLE

The metric system of measurement uses:
meters for length
grams for mass (weight at sea level)
liters for volume

To convert English measurements to metric,
or metric to English:

1 inch = 2.54 centimeters	1 centimeter = 0.3937 inches
1 foot = 0.305 meters	1 meter = 39.37 inches
1 yard = 0.914 meters	1 kilometer = 0.621 miles
1 mile = 1.609 kilometers	1 gram = 0.035 ounce
1 pound = 0.454 kilograms	1 kilogram = 2.20 pounds
1 quart = 0.946 liter	1 liter = 1.06 quarts
1 U.S. ton = .907 metric tons	1 metric ton = 1.103 U.S. tons

1 | PITTSBURGH

More than 1,600 kilometers out in space, Pittsburgh orbits earth. The spacecraft is passing over the West Coast of the United States.

It is early evening in San Francisco. The time is 6:45.

Aboard the spacecraft, local time has no meaning. Pittsburgh hurtles eastward sixteen times faster than the earth turns. The dawn of a new "day" comes to it once every two hours. It crosses the continent in eleven minutes. Below, on the East Coast, the clocks read 9:56. But the view from the ship is far wider both in space and in time.

Somewhere over the Atlantic, Pittsburgh catches up with midnight. For a brief moment, earth lies directly between the spacecraft and the sun. The crew sees the entire night side of their home planet.

The sea is dark. So are large parts of the land. Only

where there are great clusters of cities and suburbs is the land stippled with light. A solid ribbon of light outlines the East Coast of the United States, from the tip of Florida to Boston.

Earth itself is rimmed in a fiery glow. This is the air around the world's edge, lit by the rays of the hidden sun. To the west the red glow marks a Pacific sunset; to the east, dawn over Moscow and Cairo. At one glance, Pittsburgh's crew can take in the breadth of half a world and the span of half a day.

To Pittsburgh's other side, space blazes with stars, thousands more than are visible from earth. They do not twinkle, but shine with a hard, steady light. Their colors are bolder than they appear to earthbound sky watchers. From the red of Betelgeuse and Antares to the blue of Rigel and Sirius, a rainbow of starlight burns against the black sky.

But it is not to the stars that the scientists aboard Pittsburgh turn their gaze. Nor is it against the glare of the sun that they put on dark goggles as the spacecraft races toward another dawn.

They are watching a ball of white-hot molten metal. Like a tiny sun, the ball floats in the heart of an electric furnace. There are many furnaces on Pittsburgh. It is a factory in space.

Is this science fiction? No. It is science fact.

True, Pittsburgh does not exist. Not yet. But spacecraft with electric furnaces and other manufacturing equipment have circled earth.

And materials have been manufactured in space: self-lubricating alloys for auto, plane, and rocket engines; perfect crystals for use in electronic circuits; and even a lifesaving enzyme made by a culture of human kidney cells taken up into space. The enzyme was made on the Apollo-Soyuz flight in 1975. This was a joint flight by the United States and the Soviet Union.

Six years after the experiments aboard Soyuz, the next big step toward putting factories into space was taken. On April 12, 1981, the United States space shuttle Columbia blasted off on the first test flight of the shuttle system.

The shuttle is about as big as a medium-sized jetliner. But the shuttle is not an airliner. It is half spacecraft, half glider—a rocket with wings.

Once in orbit, the shuttle uses its rocket engines to maneuver as a spacecraft. On its return to earth, the shuttle reenters the atmosphere and uses its wings to coast to a landing. Unlike earlier spacecraft, the shuttle is reusable.

The shuttle is important for manufacturing in space. It is not a factory in space—but it offers a service no factory can be without. That service is transportation.

The shuttle can put factories into orbit around the earth. It can bring the raw materials to the factory and return the finished products to earth.

By July 1982, the four test flights of the shuttle were finished. The shuttle was ready to go to work as a space transportation system. And on the fourth and final test flight, astronauts Thomas K. Mattingly and Henry W. Hartsfield made some important experiments in manufacturing in space. These experiments produced certain kinds of medicine 290 kilometers from earth. They were much purer than any earth-made samples of the same medicines.

These experiments were paid for and designed by Johnson & Johnson, a United States company that makes a number of different medicines and drugs, as well as other health aids.

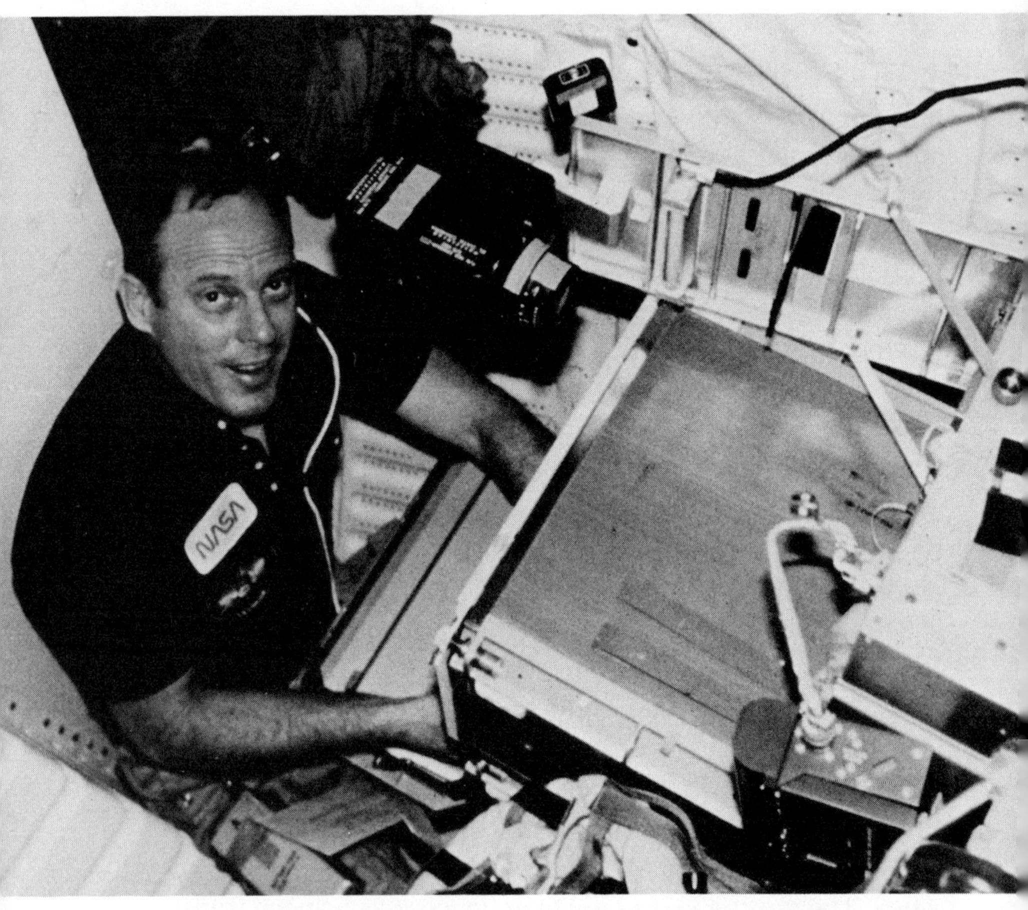

Astronaut Jack R. Lousma runs an experiment that tests the production of medicines in space aboard the Columbia space shuttle. NASA

By 1987, Johnson & Johnson plans to use the shuttle transportation system to put automated medical factories into orbit. The factories, as now planned, will be disc shaped, about 3.5 meters in diameter and one meter thick. Several of them will fit easily into the shuttle's cargo bay, leaving plenty of room for other freight. The cargo bay is 18 meters long and 4.6 meters wide.

Once the factories are in orbit, the shuttle will visit them about four times a year, bringing up raw materials and returning the medicines to earth. The factories can even be brought aboard the shuttle for repairs, if necessary.

But before that happens, scientists will have learned a lot about space manufacturing. They will learn in space, aboard earlier shuttle flights. These flights will carry Spacelab.

Spacelab is a manned scientific laboratory made to fit snugly inside the cargo bay of the shuttle. One part of Spacelab is a sealed, airtight cabin. There, several scientists can work on different experiments at the same time. The cabin is 5.5 meters long and 2.7 meters wide. Because the cabin is sealed, the scientists do not have to wear space suits. This makes it easy to use the control panels and lab equipment that line both walls of Spacelab.

Using the control panels, scientists will operate experiments inside the cabin. They will also be able to use the panels to control instruments and experiments in another part of Spacelab, toward the rear of the cargo bay. This part of Spacelab will be exposed to outer space when the cargo bay doors are open. The vacuum of space is important for some manufacturing experiments. For example, such metals as aluminum, tin, zinc, copper, nickel, and iron can be welded without heat in a vacuum. Simply touching the metals together welds them firmly together.

Two Spacelabs were built in Europe. The first Spacelab flight is scheduled for October 1983. For one week, European and American scientists will work on seventy-seven different experiments, half of them chosen by the Europeans and half of them by the Americans. By 1985, two airtight Spacelab units joined together will be carried into space by the shuttle. All the experiments on this Spacelab mission will concern manufacturing in space.

Where will these experiments lead? Will a space factory like Pittsburgh ever orbit earth? Perhaps. We are still learning what we can do in space and how to do it best. In the learning, plans will change. New ways of making things will be found. The future that holds the Pittsburgh may never come to pass.

But manufacturing in space will.

2 SPACE MEDICINE

By the end of the 1980s, if not before, some products may be labeled *Made in space.*

What kinds of products? Among the most likely are medicines—lifesaving medicines that can help in the treatment of diseases such as diabetes, anemia, and abnormal blood clotting.

Johnson & Johnson has already spent millions of dollars to test the manufacture of medicines in space. The medicines the company wants to try producing in space are *biologicals.* Biologicals are substances made by living cells that have medical uses. In one sense, living cells are microscopic chemical factories that create many thousands of different substances. All are part of the life of the cell. Some help digest the cell's food. Others make energy to keep the cell running. Still others defend the cell against invading germs or help repair the damage done by disease. So it is not surprising that many things made by living cells can be used as medicines.

Johnson & Johnson has been making biologicals on earth for many years. But some biologicals are very expensive to manufacture. They are so expensive that few people can afford them.

For example, some cells produce substances that help blood to clot when it is exposed to air. These are the substances that allow blood to clot after an injury like a cut or a scrape. Large numbers of cells clump together and stick to the broken ends of blood vessels. This stops the bleeding.

In some people, blood flowing from cuts and bruises fails to clot. These people are victims of an inherited disease called *hemophilia.* Hemophilia affects only certain boys and men. Their bodies cannot construct a key blood-clotting substance called antihemophilia factor (AHF).

Hemophiliacs may bleed for a long time from slight cuts or bruises. Bandaging with special foams can help stop the bleeding. The foams are usually made of *fibrin.* Fibrin is a sticky, threadlike substance that forms in normal blood when it clots.

Hemophiliacs often suffer from internal bleeding. This can be very dangerous and extremely painful. And of course, internal bleeding can't be treated by bandaging.

The only real protection for hemophiliacs is a large daily dose of AHF. AHF is made in the body cells of people who do not have hemophilia. Before it can be used as a medicine for hemophiliacs, it must be purified. This means it must be separated from the thousands of other substances in the cell fluid.

These substances float in the cell fluid. Like the ingredients of a complicated soup, they are made up of particles of many sizes and shapes. The particles are microscopic in size.

All the particles of any one substance—AHF, for instance—are the same size and shape. They differ in size and shape from particles of other substances.

The particles in the cell "soup" are also electrically charged. There are two kinds of charges, negative and positive. Charges of the same kind—negative and negative, positive and positive—push each other apart. Opposite charges attract.

So the particles may have either a positive or a negative charge. But the *amount* of charge on a particle can vary widely. In fact, each particle of any one substance carries exactly the same charge as every other particle of that substance. And the amount is different for each substance.

Because of this, an electric current can help separate useful biologicals out of the soup. Two wires are placed in the fluid. The wires are hooked up to a source of electrical power.

When the power is turned on, one wire becomes negatively charged, the other positive. The positively charged particles in the cell soup move toward the negative wire. The negatively charged particles move toward the positive wire.

The larger the amount of charge on a particle, the faster it moves. And, as we know, particles of different substances have different amounts of charge. So the particles of each substance move at a different pace.

Like runners in a race, the faster moving particles separate from the slower ones. Gradually, the current separates out the different ingredients of the cell soup.

This process is called *electrophoresis*. This is a scientific word that comes from two ancient Greek words. It means "carrying by electricity." And that's just what happens: The particles in the cell fluid are carried along and separated by electricity.

If electricity were the only force acting on the particles, electrophoresis would work very well. But gravity is working on the particles, too. And that makes trouble.

Why? Suppose a scientist is trying to separate two substances in the cellular soup. One is made up of particles that have a large positive charge. To remember that, we can call those particles *L* particles.

The other substance is made up of particles that have a small positive charge. We'll call these particles *S* particles.

Both L and S particles move toward the negative electrode. The L particles move much faster and soon separate completely from the S particles.

But now, suppose that the L particles are heavy, and the S particles light. What happens?

The electric current tends to carry the L particles along more rapidly than the S particles. But the heavier weight of the L particles slows them down. By contrast, the light S particles are more easily moved by the current.

So the two substances are not completely separated by electrophoresis. They remain mixed.

That's something like what happens when scientists try to separate AHF out of the cell fluid. There's another problem, too. The scientists aren't trying to separate AHF from just one other substance but from thousands of different substances.

The work of separation is slow and expensive. The final product is AHF still mixed with small amounts of many other substances. The impure AHF does not work well as a medicine for hemophilia. And the impurities may have side effects that make people taking AHF sick.

The problems of separating other biologicals by electrophoresis are the same—on earth. In space, however, electrophoresis should work much better.

The reason is weightlessness.

On earth, people sometimes are weightless for a very short time. They are weightless because they are in free fall.

Let's see what that means. Many years ago, an airplane hit the side of the Empire State Building in New York City. A part of the plane ripped through some elevator cables about eighty stories up. Elevators, with passengers aboard, began to plunge toward the bottom of their shafts. They fell faster and faster, under the steady pull of gravity.

That wild plunge didn't last long. Automatic emergency brakes gripped the guide rails, slowing the elevators down. At the bottom of the shafts, powerful springs cushioned the fall. The passengers were bruised, badly frightened—but safe.

Before the brakes took hold, those passengers were weightless. The elevators were in free fall. Like an astronaut in orbit, the passengers experienced zero gravity.

The passengers were falling. But the elevators were falling at the same speed. So anything a passenger dropped would never reach the floor. Everything was weightless . . . until the emergency brakes took hold. They gripped the guide rails and slowed down the plunge of the elevators. They were no longer in free fall. There was another force—the force produced by the grip of the brakes—resisting the pull of gravity.

At once, the situation changed. Things dropped abruptly to the elevator floor. People felt their feet pressing against the floor once more. Weight returned.

Free fall never lasts long on earth. It was just a few seconds from the time the elevator cables broke until the brakes slowed the fall.

But in space it is possible to be in free fall for as long as you want and never worry about "hitting bottom." The scientists and crew of a spacecraft orbiting earth are in perpetual free fall. And so, they are weightless.

They are not weightless because they have escaped earth's gravity. It is the gravity of our planet that holds the spacecraft in orbit.

In fact, the tug of gravity on a spacecraft 320 kilometers from earth is nearly as strong as it is at earth's surface.

The spacecraft is hurtling along in its orbit at a speed of more than 27,000 kilometers an hour. Without gravity, the spacecraft would move steadily farther from earth in a straight line. But as the ship's speed carries it *forward,* gravity tugs it *downward.* The ship falls downward just enough to stay the same distance from earth as it moves forward. So its orbit is a curving path around earth. A spacecraft in orbit is in free fall *around* earth.

In fact, a spacecraft in space is always in free fall, so long as its rockets are not firing. And for those aboard, the effects of gravity don't exist.

So weightlessness or zero gravity is a natural part of space travel. Actually, scientists prefer to use the word *microgravity* to describe what it's like aboard a spacecraft in free fall. That's because the spacecraft itself has gravity. So do every person and bit of matter in it. They all exert a pull on one another. But the sum of all these pulls is very feeble.

A man who weighs 90 kilograms on earth would weigh a tiny fraction of a gram in space.

Microgravity is a big part of what makes manufacturing in space possible. For years, scientists predicted that electrophoresis would work better in space than on earth. Then, in 1975, the prediction came true.

In that year, the United States and the Soviet Union cooperated in a joint space flight. On July 15, 1975, an American Apollo spacecraft docked with a Soviet Soyuz spacecraft in earth orbit. The two ships orbited earth, and the Soviet and American astronauts visited back and forth.

During the time the spacecraft were linked together, some of the first experiments in space manufacturing were tried. One was an experiment in electrophoresis.

For this experiment, scientists had carried a culture of human kidney cells aboard the Apollo spacecraft during the Apollo-Soyuz mission. About 5 percent of human kidney cells make a substance called *urokinase*. Urokinase dissolves blood clots that sometimes form in circulating blood.

Unlike the blood clots that begin to form as soon as blood is exposed to air, the clots that form in circulating blood are dangerous. They can cause strokes or sudden heart failure. In the United States alone, about fifty thousand people die each year from such blood clots.

To save these people would take about five hundred thousand doses of urokinase a year. A single dose now costs more than a thousand dollars. And only small amounts of the medicine are produced.

Why? The 5 percent of kidney cells that produce urokinase must be separated from the 95 percent that don't. Electrophoresis can separate different kinds of whole cells, just as it can separate substances within cells.

But the problems are much the same in each case. The process is slow and expensive. The separation is far from perfect.

On the joint Apollo-Soyuz flight, though, electrophoresis worked much better. In microgravity, electrophoresis produced a culture of urokinase-making cells much purer than any culture ever made on earth. These cells made seven times more urokinase than the best earth-made cultures.

The Apollo-Soyuz electrophoresis experiment was a success. It showed that, thanks to microgravity, space is a better place than earth for making certain medicines. Space is also a good place for making a vital part of all computers—and of video space games.

3 CHIPS, FIBERS, AND WALLS MADE OF ENERGY

Someday soon, signs like this one may appear in video game arcades:

PLAY
SPACE BANDITS!
THE FIRST SPACE GAME MADE IN SPACE
BY SPACE NUTS FOR SPACE NUTS!

And why not? The heart of any video game is a tiny chip like the one shown on page 18. That chip is a miniature computer—a microchip.

Computers used to be a little bigger. Take ENIAC, for instance.

ENIAC was the first electronic computer. It was invented in 1946 by a professor and a graduate student at the University of Pennsylvania.

ENIAC, the first electronic computer, filled an entire room.
THE MOORE SCHOOL OF ELECTRICAL ENGINEERING, UNIVERSITY OF PENNSYLVANIA

The proud parents named their baby ENIAC, which was short for *E*lectrical *N*umerical *I*ntegrator *a*nd *C*omputer.

ENIAC was a rather large baby. It was cranky. While it did not need diapers, it certainly had to be pampered. When the inventors wanted to give ENIAC orders, they had to get into it and change the wiring by hand.

So it was just as well that ENIAC filled a room and weighed 27 metric tons. At least that made it easy to get at.

ENIAC was also a very greedy baby. It used up huge amounts of electric power and wasted most of it as heat. What produced the heat were the red-hot filaments inside the nineteen thousand vacuum tubes that did most of ENIAC's work. The tubes overheated so often that one of the nineteen thousand blew out about every seven and a half minutes.

Yet ENIAC was a marvel in its day. It could do five thousand calculations a second. It could do about as much as—you guessed it—the microchip that steers the ghosts chasing Pac-Man—or Ms. Pac-Man, a new and harder version.

As much, but not nearly as fast. Microchips like Pac-Man's can do millions of calculations in a second. Yet thousands of such chips could fit inside just one of ENIAC's vacuum tubes. The circuits in a chip are more complex than all of ENIAC's room-filling wiring—but so small you'd need a high-powered microscope to see them.

Because of the microchip, computers are getting more powerful for their size and cheaper at the same time. By 1990, inexpensive pocket calculators will be able to do all that large computers ever could—and this includes large computers that in the late 1970s and early 1980s cost millions of dollars.

What's more, microchips can be made better—and in the long run more cheaply—in space. The microchip industry will be one of the first to use space as a good place to work.

Without high-powered computers, space travel would be nearly impossible. Human calculators would take far too long to wade through all the complicated mathematics needed to make space flight safe and accurate. And now it turns out that space is a better place than earth for making the high-powered computers' vital parts. The future of space travel and the future of microchips may well depend on each other.

Why is this true? Let's trace, briefly, the story of the incredible shrinking computers.

All computers use the binary number system. This sys-

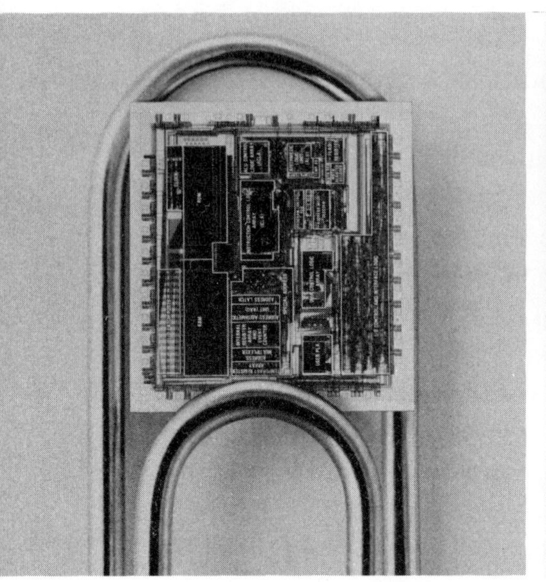

A one-chip computer (left) is compared to a standard-size paper clip; a data processing chip shown next to a shirt button (right) contains nearly six hundred tiny, fast switches.
COURTESY OF BELL LABORATORIES

tem has only two symbols, zero and one. It is a binary (base two) system.

Our number system is a decimal (base ten) system which uses ten different symbols: 0, 1, 2, 3, 4, 5, 6, 7, 8, and 9.

That difference aside, the two systems are mathematically equal. Any number can be written in either system. Any mathematical operation can be done in either system.

For people, the binary system would be very awkward. Numbers grow rapidly in size. Our familiar 8 becomes 1000 in the binary system. One hundred becomes 1100100.

For a computer that handles millions of calculations a second, big numbers are not a problem. And the binary system is natural for computers. That's because all computers work by electrical switching.

In ENIAC, each vacuum tube worked like an electric switch. A switch is either on or off. It lets a current flow or it blocks the current.

The tubes worked in the same way. Whether the tubes blocked a current or let it flow depended on many things: for example, on how the tubes were wired together, and on the changing direction of current flow in the circuits.

Within ENIAC, thousands of switches were going on and off thousands of times a second. The changing patterns made by the switches expressed numbers in the binary system. Each "off" switch stood for a zero, each "on" for a one.

Today's computers work in much the same way, but the switches and the wiring are now microscopic in size. Millions of switches are going on and off millions of times a second. In the near future, this speed will rise to billions and trillions of times a second.

Numbers, letters, images, sounds, and speech can be turned into the on-and-off code of the modern computer. The computer can process symbols and signals in this code according to its programs. It can turn the processed material back into printed information or pictures on a television screen, or into sounds. Or the computer can use the information for direct control of a maneuver. That's what happens when a computer guides the shuttle to a safe landing.

The microchip that makes all of this possible is usually a tiny crystal of pure silicon. Crystals of a few other substances are also used to produce microchips. As we shall see, more perfect crystals of any sort can be made in space. What is true of silicon crystals made in space is true of these other crystals. Silicon is a cheap and plentiful element. Most of the sand and rock on earth is made up of silicon dioxide. That's true of the other rocky planets in the solar system, too—Mercury, Venus, and Mars. It's also true of many of the moons of the solar system, and probably of many asteroids.

The silicon is "doped" with bits of certain other elements. These bits of impurities are imbedded in the silicon. The atoms of impurities and the surrounding silicon atoms work together as an electrical switch. The only moving parts are the electrical charges around the atoms.

Thousands of such switches are built into one chip. They are joined by complex circuits. The circuits are designed and tested by engineers to fit the computer's job. The circuit design is reduced to microchip size on film. Then the circuit is printed photographically onto the chip.

Chips like this are mass-produced from a cylinder-shaped crystal of silicon about 18 centimeters long and 13 centimeters in diameter. The big crystal is sliced into "wafers" about half a millimeter thick. Three such wafers are not quite 1.59 millimeters thick. Each wafer can be made into more than a hundred identically wired and doped chips.

This can be done because big crystals are made up of little crystals, which in turn are made up of even littler crystals—almost down to the level of atoms. The molecules or atoms of a crystal are arranged in exactly repeating geometric patterns.

With crystals, "a chip off the old block" is really a miniature twin of the old block. Not just in looks or makeup, but in electrical properties, too. And that's one important reason why silicon crystals are so useful in making microchips.

Crystals must be grown. Silicon crystals are made by melting a batch of silicon in a crucible. That takes a lot of heat, since silicon melts at 1420 degrees Celsius (nearly

2600 degrees Fahrenheit). Then a small "seed" crystal of silicon is suspended in the melted silicon.

The seed is colder than the melted silicon. So new layers of crystal form around it. The crystal is allowed to grow until it is big enough to be sliced into wafers.

But earth-grown crystals are never perfect. They have tiny flaws scattered through them. Each flaw is a break in the pattern of the crystal. When these flaws become part of a microchip, there are problems. There are breaks and changes in the flow of current through the chip. Either the chip doesn't do its job right to begin with, or it may fail in use.

Such flaws are very common. We've mentioned that over a hundred chips can be made from one wafer. But out of each hundred chips, on the average, only twenty-one will be good enough to become part of a computer. The rest must be thrown away.

Why do the crystals have so many flaws? The chief villain is gravity.

As we've seen, the silicon crystal grows in a pool of melted silicon. But that pool is far from calm and still. The same heat that melts the silicon stirs it up. Within the pool, liquid silicon is constantly flowing up and down.

You can watch this same kind of flow in a pot of soup warming up on the stove. That's what makes the solid bits of food in the soup keep churning.

Churning is fine for soup. But it's not good for growing crystals. It causes them to grow unevenly, with many flaws.

On earth this can't be avoided. Whenever a fluid is heated, some parts of the fluid get hotter than others. The hotter parts are lighter than the cooler parts of the fluid. So the hot parts rise, then sink again as they cool.

In space that can't happen. There is no difference in

weight between the hotter and cooler parts of a fluid—because weight doesn't exist. As early as 1973, a flawless crystal was manufactured in a space experiment. The experiment was repeated on the joint Apollo-Soyuz flight two years later. In both cases, the crystals were perfect. Their electrical properties were exactly the same from one end of the crystal to the other. Even the best earth-grown crystals don't come up to this standard.

These experiments were just the beginning. Now the National Aeronautics and Space Administration (NASA) and a company called Microgravity Research Associates are planning to put automated crystal-growing factories in space. Such factories may be in orbit by 1990 or even earlier. They would make crystals of silicon and other substances used in microchips.

One scheme is to manufacture long ribbons of silicon crystal. The silicon would be melted in a furnace using concentrated heat from the sun. The cooling silicon would be drawn out into ribbons and wound onto turning reels, much as movie film is wound onto the reels of a projector. The moving parts of the space factory would get their power from solar cells. These are panels of cells that turn sunlight directly into electric power.

Such cells often use silicon as a working substance. Silicon produces electricity when light energy strikes it. And the silicon ribbon made in space would be ideal for use in solar cells as well! So once again, we have a product that is useful in space and that can be best made in space.

Space manufacturing may also lead to better and cheaper telephone service here on earth. The "wires" that will make that possible are made of glass, not metal. Each is only half as thick as a human hair.

A cutaway view of a telephone cable shows twenty-two coaxial tubes. COURTESY OF BELL LABORATORIES

The old-fashioned telephone cable is packed with 1,800 copper wires, forming a bundle about 11 centimeters thick. It can carry 900 phone conversations at the same time.

The new phone cables are made up of bundles of glass fibers wrapped in a plastic sheath. Strands of steel are wound into the plastic to give the cable strength. These cables are only 1.3 centimeters thick. They contain 144 fibers compared to the 1,800 wires in the old-fashioned cable. Yet they can transmit 500,000 conversations at the same time—almost 600 times as many as the old-fashioned cable wired with metal.

How is this possible? The secret is in the way the messages are carried.

In the old-fashioned cable, the messages are transmitted by electricity passing through the copper wires. The speaker's voice is translated into an electric current. Changes in sound are turned into changes in the current. At the receiving end, the changes in current are translated back into sound again. It takes one wire to carry each side of the conversation. So the 1,800 wires in the cable can carry 900 conversations at the same time.

But glass fibers do not carry electricity. They carry pulses of light—pulses that go on and off millions of times a second. The human voice is translated into these pulses. As you may have guessed, this is the same on-and-off binary code that makes computers work.

The device that makes the pulses of light is a laser, smaller than a grain of salt. Lasers produce strong, narrow beams of light. Unlike the beam from a flashlight or searchlight, a laser beam can travel millions of kilometers without spreading much. A laser beam can be flashed from earth to the moon with pinpoint accuracy.

A microchip is used to code the voice into pulses of light at one end and to translate the pulses back into sound at the receiving end. In effect, the voice is chopped up into millions of pulses of light each second.

A wire cable (top) dwarfs the optical fiber cable. But the fiber cable can carry 500,000 conversations at once. That is almost 600 times the capacity of a typical old-fashioned wire cable. ▶
COURTESY OF BELL LABORATORIES

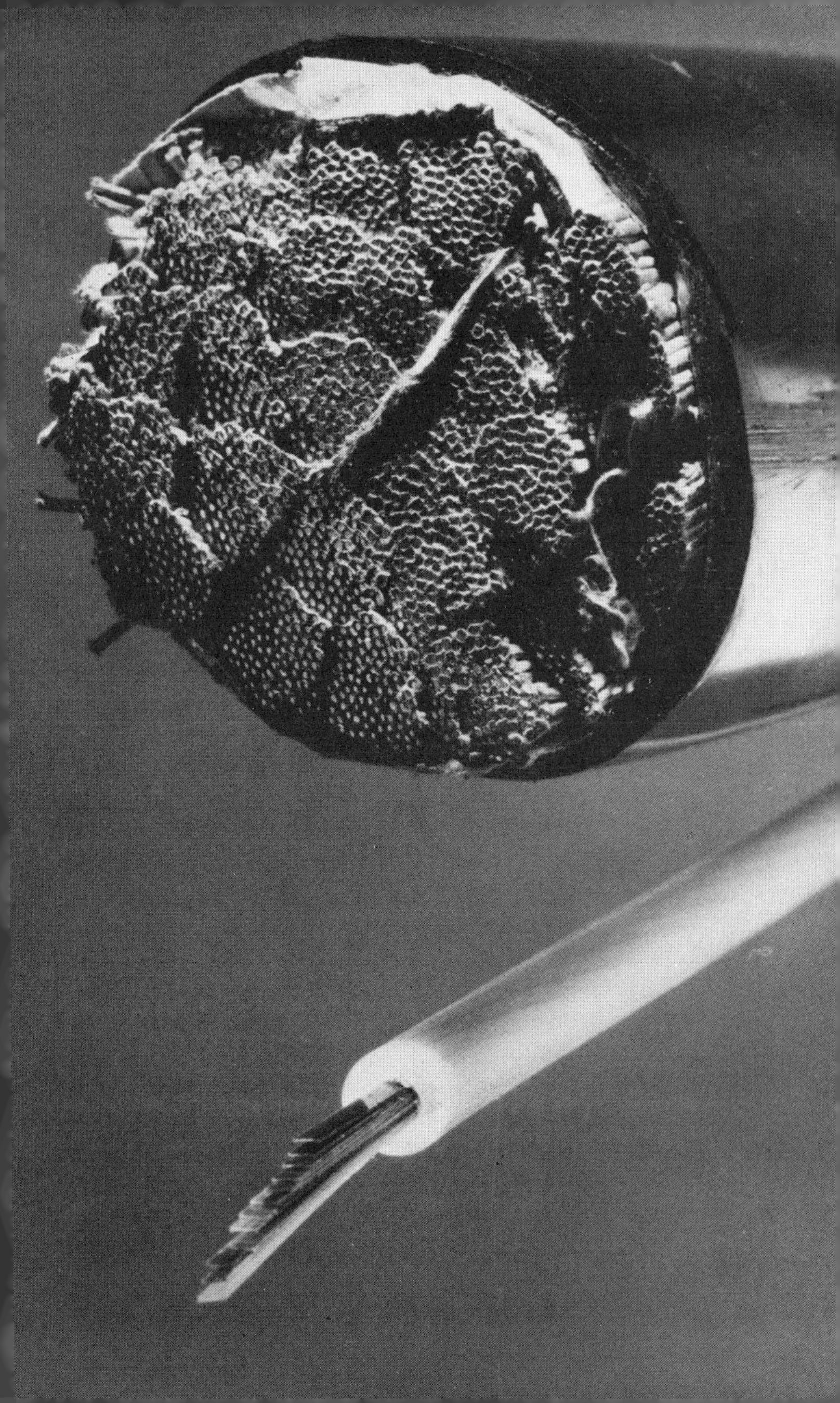

Chopping up a voice into so many tiny bits of signals per second leaves a lot of spaces where nothing is happening—where no voice sounds are being made. There are pauses for breath, pauses between words, and pauses at the ends of sentences, for example.

For a circuit and laser beam working millions of times a second, these spaces are huge. Other messages can be coded into them. In this way, thousands of messages can travel down one fiber and be unscrambled at the receiving end. This is true not only of voice signals but of music and television images and sound.

As early as 1976, messages were sent through a light cable for a distance of nearly seven miles. By 1984, a light cable network will join the cities of Washington, D.C., Philadelphia, New York, and Boston. It will be able to carry 80,000 simultaneous messages.

The glass fibers carrying these messages are called optical fibers. The glass in these microscopic light-carrying tubes must be as free as possible of dust, dirt, scratches, and impurities.

Here's why: Light normally travels in a straight line. But, obviously, light-carrying cables can't be laid out in straight lines. The light pulses going through the fibers must twist and turn. They must flow through the fibers as water flows through pipes.

This can be done because the walls of the fibers are mirrorlike. The light bounces back and forth from wall to wall as it flows down the "light pipe." Because the walls are nearly perfect reflectors, the light pulses lose very little energy.

Very little—but some. Small amounts of energy are lost, due to impurities in the fiber walls. This loss dims the light

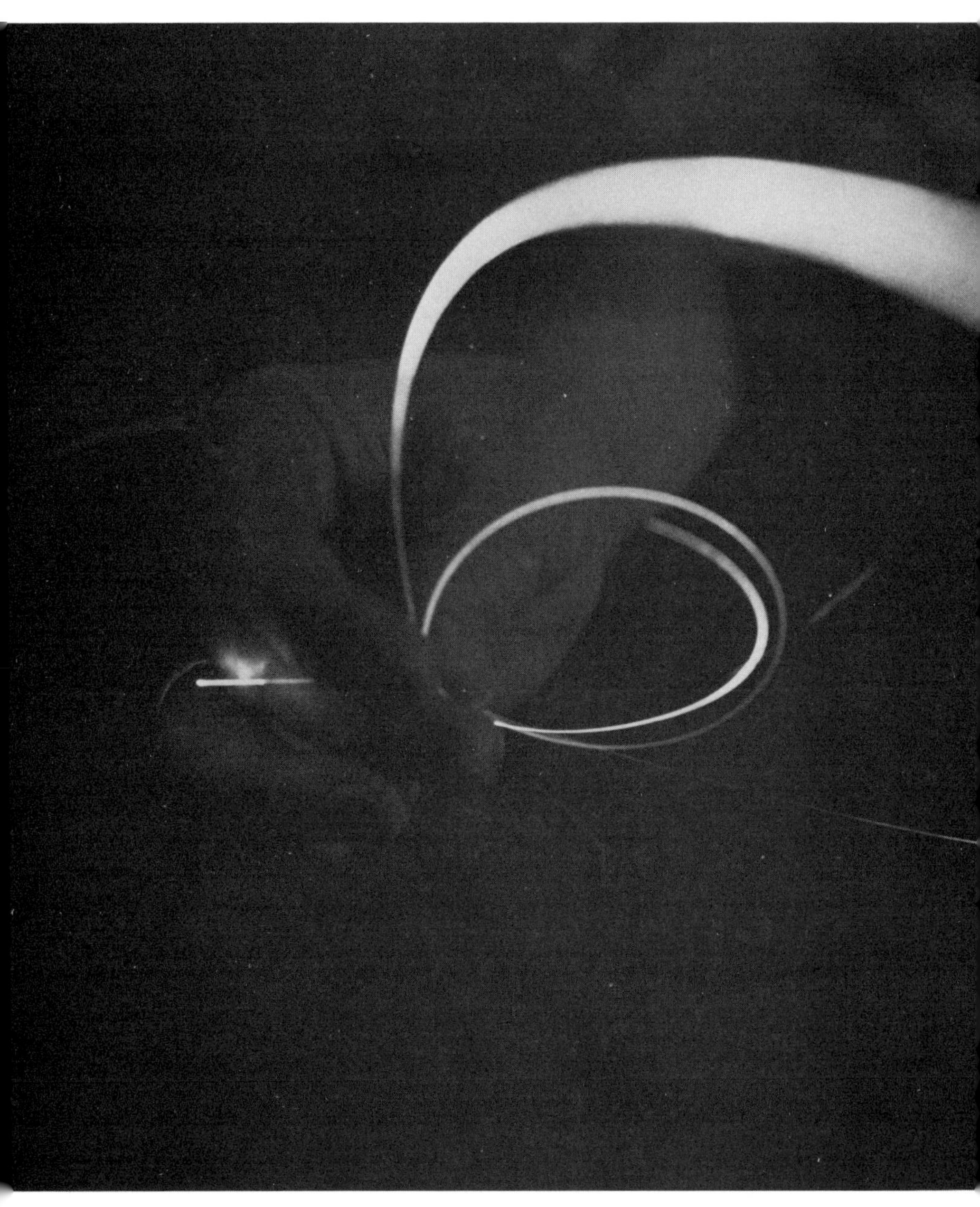

Loops of a hair-thin glass fiber, illuminated by laser light.
COURTESY OF BELL LABORATORIES

pulse traveling through the fiber. Eventually, the message could be destroyed.

To prevent this, repeater circuits boost the light signals at regular intervals along the cable. The best earth-made fibers need repeaters about every 10 kilometers. Scientists calculate that optical fibers made in space could double that distance, thus cutting the need for repeaters in half.

Once again, the major advantage of space manufacturing is microgravity. The glass used in optical fibers must first be melted. The molten glass picks up impurities from the sides of the container that holds the glass.

This is a problem with most manufacturing on earth. Many useful materials must be melted at some point in their making. The liquid must be contained in something. Because the fluid is usually very hot, it quickly dissolves impurities on the container walls. If the fluid is used to make crystals, there is more trouble. Crystals tend to form wherever there is the slightest scratch or rough spot on the container wall and will be distorted by the rough surface.

In space, however, molten liquids do not need a container. Or rather, the container doesn't have to be made of any kind of material. The weight of the liquid is practically zero. So it can be suspended in midair by small amounts of energy. The energy can be in the form of electrical or magnetic forces, or even sound waves.

Containerless operation is another reason why crystals can be made better in space, too. In fact, the silicon ribbon space factory that we've talked about will use this method. Electrical forces will hold the molten silicon in place. As the liquid cools, the same forces will pull it out into ribbon shape.

All of this may sound far out. And space factories will be far out—in space. But they are very near in time.

4 LABS AND PLATFORMS IN SPACE

If all goes well, the manufacturing of medicines in space will start around 1987. From 1987 to the end of 1988, the space shuttle will carry one factory into orbit every six months. By the end of 1988, the shuttle will have transported five space factories into orbit.

Each factory will be disc shaped and will take up about one meter in the shuttle's payload bay. Since the bay is 18 meters long, that's not much space.

The factory will be light by shuttle standards, weighing about 1,600 kilograms. The shuttle can carry a payload of 30,000 kilograms.

Once in orbit, the factories should produce medicines of much higher purity than possible on earth. The factories will be automated. On-board computers will control production. Electric power to run the computers and the factories will come from sunlight.

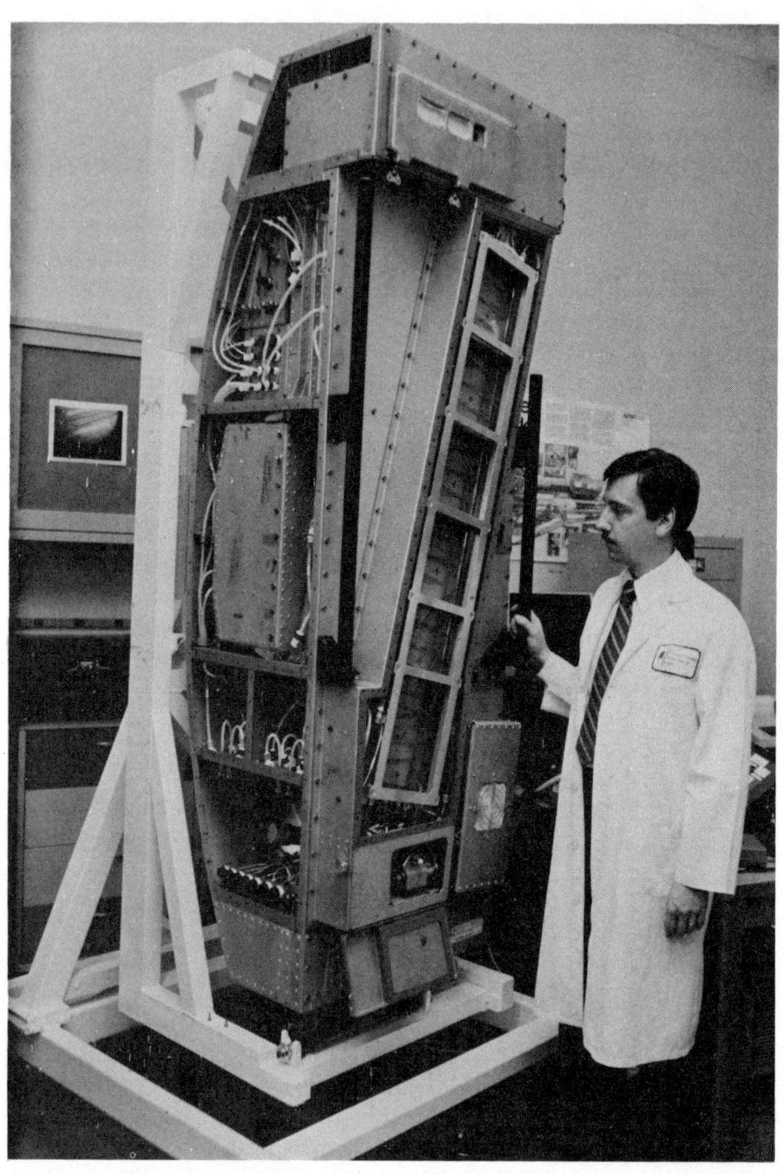

An electrophoresis device like this was tested on the sixth and seventh shuttle missions. In zero gravity, the device produced much purer samples of some biological materials than can be made on earth. This is an important step toward manufacturing certain disease-fighting medicines in space. MC DONNELL DOUGLAS CORPORATION

About four times a year, technicians will visit the orbiting factories. They'll bring new supplies of raw materials and pick up the finished products for transfer back to earth. The technicians will also check up on the computers to make sure the factories are running properly.

The factories can be small and light because the raw materials—mostly living cells or cell parts—are small and light. So are the finished products.

All this cuts down on the costs of putting the factories into orbit, keeping them supplied and operating, and getting the products back to earth. For these reasons, scientists hope that some space-manufactured medicines may eventually sell for only a tenth of the cost of the same products made on earth. So medicines may become one of the first successful space businesses.

So may microchips for computers, communication, and many other uses. Like the medical factories, microchip space factories will be run by computers and supplied regularly from earth.

As we've seen, microchip technology is changing rapidly. To keep up with these changes, the factories may be brought back to earth every four or five years for redesigning. This will be one of the shuttle's big jobs. Using a remotely controlled mechanical arm, satellites and factories can be picked up, stored in the shuttle's cargo bay, and returned to earth for major repairs.

Scientists now figure that two or three microchip factories in space could supply about half of our growing need for microchips by the end of the 1980s. If that happens, the computers running these factories may be using microchips made in the factories!

The microchip factories will be somewhat bigger and heavier than those making medicines. That's because the raw materials, such as silicon, are bulkier than cells and cell parts. Also, the process of making the chips will use more complicated equipment.

For the same reasons, making glass fibers and other glass products in space will need still heavier and more complicated factories. And still larger factories will be needed for making alloys.

Alloys are a blend of one or more metals with other metals or nonmetals. Steel, for example, is a blend of iron and carbon. The blend is stronger and harder than iron by itself. The use of alloys goes back to the Bronze Age of prehistoric times. Bronze is a blend of copper and tin with small amounts of other elements. It is easier to shape than iron and more rust resistant. Yet bronze is much tougher than the copper and other elements that go into making the alloy.

Complex alloys of many kinds are used in precision tools and machinery. Alloys form vital parts of rocket engines and satellites. Yet the basic idea behind alloys remains what it was in the Bronze Age. The idea is to change the properties of a metal by blending it with other substances—and to get just the right blend for each job.

A touch of aluminum makes bronze more corrosion resistant. A touch of tungsten makes steel super hard. The recipes for alloys would fill a fat cookbook. And more are being discovered all the time.

Some alloy recipes, however, just don't work on earth. You can't make an alloy of aluminum and lead on earth, for instance, yet such an alloy should have some very useful properties. It could be even more useful than alloys of aluminum and tin, which can be made on earth and are used to cut down friction between moving parts. Aluminum-tin alloys are used in automobiles. But aluminum-lead alloys would last longer and be self-lubricating. In theory, aluminum-lead alloys would make it possible for an auto engine to last more than 800,000 kilometers.

Why don't aluminum-lead alloy recipes work on earth? Lead is four times more dense than aluminum. And on earth that means lead is over four times heavier.

To form the alloy, the lead and aluminum must be melted and blended, then cooled again. Once melted, the aluminum floats to the top.

Stirring doesn't help. In fact, as we've seen, the heat used to melt the two metals creates a strong flow in the molten mass. But the great difference in weight tends to separate the metals still more. The molten mass becomes a molten mess and cools to a useless lump.

In space, lead is still four times denser than aluminum. But both lead and aluminum are weightless. So we should be able to make an alloy out of them.

That's been done. In one microgravity experiment, a small amount of aluminum and lead formed a smooth alloy. The experiment was done aboard an unmanned rocket flight sponsored by a group of West German companies.

The rocket was launched from Norway. It did not go into orbit, but rose to a height of 260 kilometers and fell back to earth again. A parachute brought the payload to a soft landing.

For about six minutes, the rocket was in free fall so that everything aboard was weightless. During that time, automated equipment did a number of experiments related to manufacturing in space.

These experiments were just a rehearsal for longer ones that will take place in the near future. But they showed promising results.

In one experiment, a turbine blade was manufactured.

The blade is used in jet airplane engines. It is made of a special high-strength, heat-resistant alloy. The alloy was melted and shaped within a "skin" of aluminum oxide less than a tenth of a millimeter thick. The result was a blade many times stronger than could be made on earth. Blades like these can last longer and operate at higher temperatures than blades used today. That's important, because jet engines work better and therefore use less fuel at higher temperatures.

Altogether, scientists have listed about four hundred alloys that might be manufactured better in space because of microgravity. Some of the alloys on the list have never been made at all. We will have to find out if they can be made in space and how useful they might be.

It will be some time before alloy-making factories—or glassmaking factories—are circling earth. They will be bulky and heavy. It may be easier and less expensive to send them up piece by piece and put them together in orbit.

Before that happens, scientists will have learned a lot about space manufacturing. They will have learned in space. Their first classroom will be Spacelab.

The people who designed Spacelab learned much from the problems of Skylab, the first United States space station and laboratory. Skylab was launched on May 14, 1973. It was "home" to three different teams of astronauts during 1973. After that, Skylab was unoccupied. It broke up on reentering earth's atmosphere, July 11, 1979.

Skylab had a cylindrical cabin lined with instruments and equipment all the way around. The effect of this, even on experienced astronauts, was dizzying. In the microgravity of space there is no up or down—unless it is built into the design of a place.

So Spacelab is not perfectly cylindrical. It has a flattened

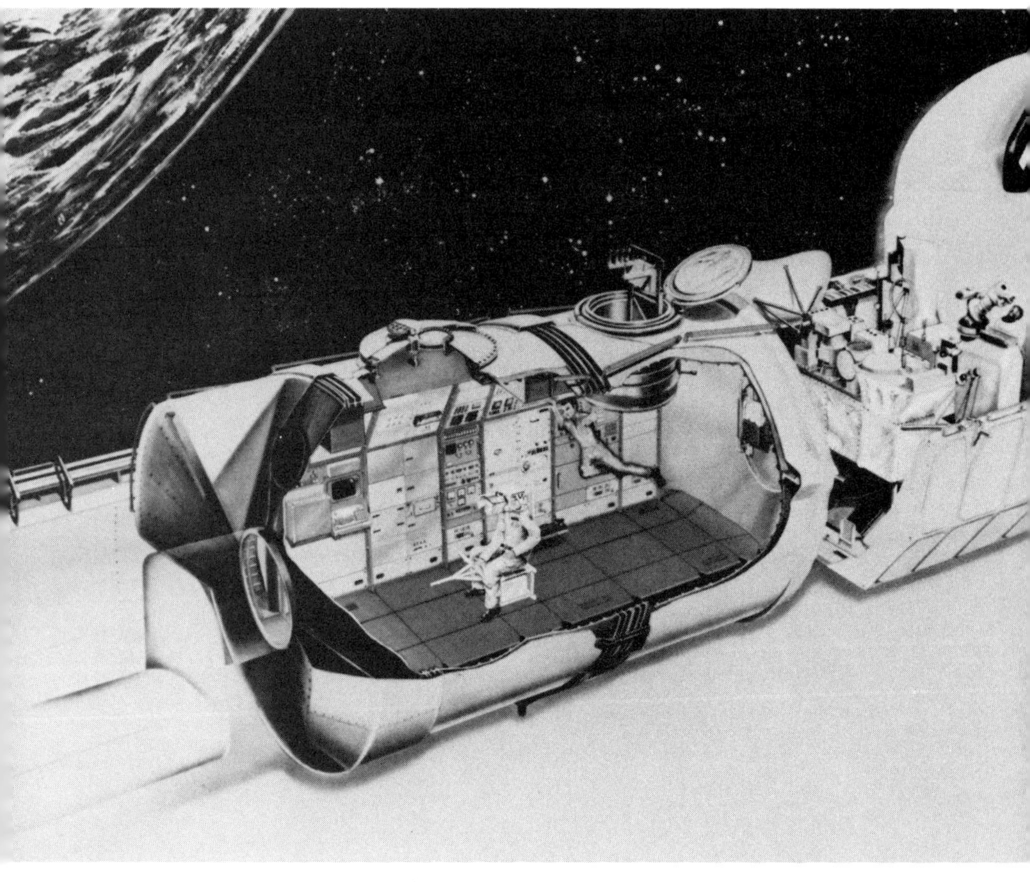

Spacelab 1 with its payload. EUROPEAN SPACE AGENCY

floor and ceiling. Only the walls are lined with equipment. There are handrails along each wall, at right angles to the floor and ceiling. That gives the scientists an added sense of "up" and "down." It also gives them a kind of hand brake. A gentle push will send a scientist floating along the length of Spacelab. Grabbing a rail will bring him or her to a stop. When working on an experiment, suction cups on the soles of the shoes anchor the experimenter to the floor.

Toward the back of Spacelab is a porthole for viewing.

Through it, scientists will be able to gaze at stars—more stars than can be seen from earth, stars with more definite colors that shine steadily without a twinkle. Or they will be able to take time to discover how beautiful a planet the earth is: the blue green of its vast oceans; the variety of its lands, from the green of forests, prairies, and jungles to the brown of deserts and the brilliant white of the polar caps; and the swirls of clouds, making the very storms and winds of our planet visible.

Spacelab's view port will have handrails all around it. Skylab's did not. Its crew found out they could not get a clear view of earth and space through the port. Without handrails to hold on to, every slight movement sent the observers shifting and drifting.

Spacelab scientists will have a more convenient, comfortable, and better place to work in than the space travelers who came before them. In turn, a good part of the research done on Spacelab missions will pave the way for space factories. Experiments will show which methods of manufacturing work and which do not.

Spacelab experiments will help answer questions such as: What ways of manufacturing crystals work best in space? What kinds of materials are best made by containerless operations? What useful new alloys can be made in microgravity? What medicines can be made better in space than on earth?

However, as a place to test plans for space factories, Spacelab has drawbacks. Instruments and experiments in Spacelab must get their electrical power from the shuttle's fuel cells. The fuel cells also supply power for the shuttle's five on-board computers and other essential equipment.

The shuttle's fuel cells can only produce power for about seven days. The cells may be improved to last a longer time. But at best, the shuttle will have to return to earth after about two weeks in space. And Spacelab will return with it.

As one scientist at NASA, Leo Zoller, points out, there are any number of ideas for manufacturing in space that need to be tested. They can only be tested in space. Many of them will take longer than a week or so to do. The products will have to be returned to earth for further testing. More materials and equipment must be brought from earth. What is needed is a permanent space station.

"We'll need to ferry hundreds of specimens back and forth [from earth to orbit]," says Dr. Zoller. "We're really looking forward to the station so we can get a lot of work done instead of one shot at a time."

Dr. Zoller is talking about the idea of a space station—a permanent base in space. The first step to the space station might be the space platform now being planned at the Marshall Space Flight Center in Huntsville, Alabama. The space platform has a box-shaped central module. This holds radios and antennas for sending and receiving messages from earth, and attitude-control equipment. The attitude-control instruments keep the platform pointed in the right direction in space.

The platform will be carried into orbit aboard the shuttle. Once in space, long metal booms will extend out on either side of the central module. When fully extended, the booms look like long, thin wings. They hold solar power cells. The cells convert sunlight directly into electric power.

The platform will be an unmanned power station. It will be a place to attach experiments and plug them in. The ex-

A space station concept designed by the Marshall Flight Center that includes stations for servicing unmanned scientific platforms. NASA

periments will be bolted onto U-shaped frames called *pallets.* Such frames were successfully tested on the second, third, and fourth shuttle flights.

The shuttle will carry the experiments into orbit mounted on pallets. They will be attached to the platform at special berthing ports.

The space platform simply makes use of another one of

A space platform designed by McDonnell Douglas for production of drugs in orbit. At lower right, a shuttle orbiter is docked with one of the free-flyer units during a resupply mission. MC DONNELL DOUGLAS CORPORATION

the advantages of space—the practically unlimited supply of energy. It comes from the solar system's main power plant—the sun. And not only is it free, it is also steady and reliable. It can't be interrupted by power outages, downed wires, broken-down generators, clouds, or storms. Experiments can be operated for weeks or months at a time on the space platform.

A huge satellite power system like this could be built in orbit. Orbiting platforms would harness the sun's energy and transmit it to microwave receivers on earth. The structure would be covered with thousands of acres of solar cells which convert sunlight to electricity. BOEING AEROSPACE COMPANY

Sooner or later, though, people will be needed to control and check on some of the more complex experiments. The space platform will then become a manned space station.

This may be done by joining two modified Spacelab sections together and connecting them to the platform's central module. Each section will have its own living quarters and lab space and could house a team of four scientists. After several months, the shuttle will bring up new supplies of oxygen, food, and equipment. Teams will be rotated so that no team spends too long a time in space.

Space shuttle astronauts erect a large space structure of the future. An automated beam builder aboard the space shuttle's cargo bay produces girders or beams from coils of lightweight metal plate. NASA

This possible design for a space station, conceived by Rockwell International, includes two solar panels to provide power; modules for command, living quarters, and experiments; and a shuttle orbiter. NASA

An advanced space operations center designed by Boeing. The shuttle unloads modules that include living and command control quarters; warehouses for food and water; and service areas. Solar panels provide power. NASA

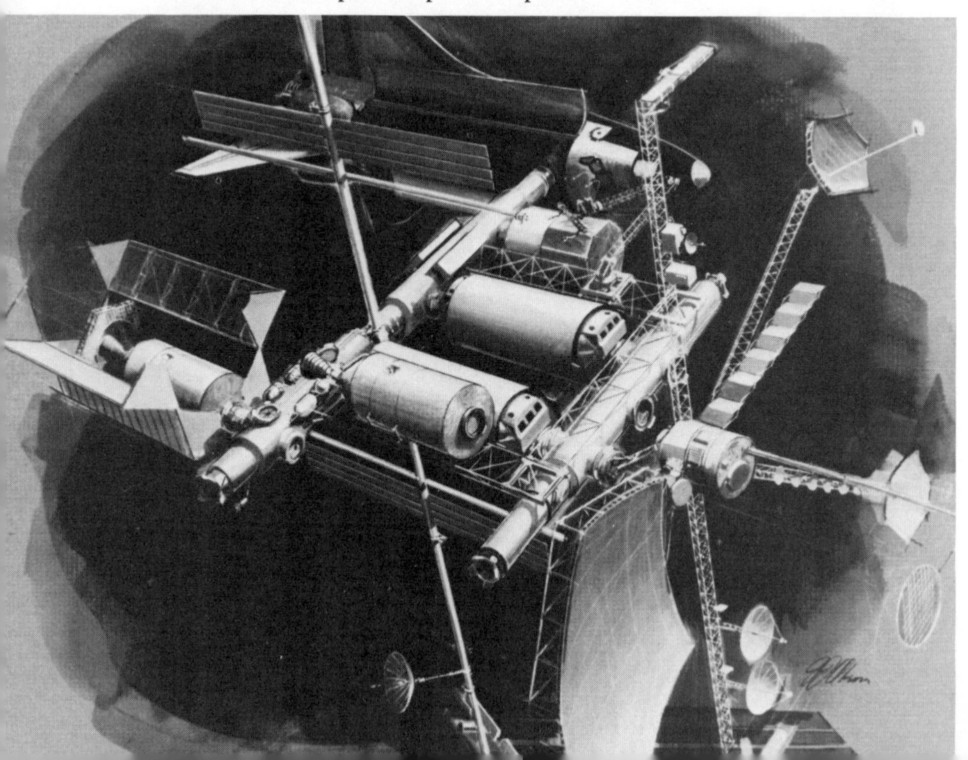

A concept for an orbiting space station designed by Lockheed Missiles and Space Company. The station features artificial gravity living quarters and control center; three modules for scientific, commercial, and other applications; and three open beam platforms. NASA

We already know how to supply food, water and air for crews on a long space flight. Supplying the crew of a space platform will be done in much the same way:

Oxygen is brought up in tanks. The oxygen is compressed and cooled until it becomes a liquid. Liquid oxygen can be stored in much lighter and smaller tanks than the same amount of oxygen gas.

Food is freeze-dried to remove almost all its water. The dry powder is then sealed in small packets. Since water makes up over 75 percent of the weight of most foods, freeze-dried food is light and easily carried into space.

A day's supply of freeze-dried food for one space traveler weighs only about half a kilogram. Nearly 2 kilograms of water must be added to this before it can be eaten.

People need more water by weight than they do food and air put together. It would be very costly to keep resupplying a spacecraft with water from earth. Instead, water aboard the spacecraft is recycled. It is used again and again. That includes the water people lose from the air they breathe out and from sweat. This water passes into the air in the form of a gas—water vapor. Water vapor turns back into droplets of liquid water when it touches a cold surface. This is what happens when water drops collect on the outside of a glass that holds an iced drink.

All waste water is cleaned, sterilized, and used again. The spacecraft is sealed airtight against the vacuum of outer space. So little or no water is lost.

Once a space station is in orbit, the work of finding the best design for space factories will go ahead much faster. But when the final design of a particular kind of space factory is drawn up, what will it look like?

That remains to be seen. What is almost certain is that space factories will be automated. Transporting people into space just to have them do routine jobs is too expensive. People need air, water, warmth, and food. They must be shielded from dangerous solar radiation and from poisons produced as by-products of manufacturing.

Machines don't need any of this. What's more, a computer operating a factory doesn't make mistakes out of boredom or because it is tired or upset. Within limits, a computer can switch from one routine to another. It can make logical choices in an emergency.

Computer-operated spacecraft have landed on Venus, where no person could survive for a second. Computers have guided the spacecraft to a soft landing on the planet. They have operated cameras to take pictures of the surface.

Similar landings have been made on Mars. Samples of soil were scooped up and carried to a miniature laboratory inside the spacecraft for testing.

Of course, much of this activity was controlled by radio signals from earth. But computers exist that can do this sort of job on their own.

Humans have a role to play in space. That role is one of discovery and invention. Most often a discovery is an accident that grabs the imagination.

On the third Skylab mission, for example, astronaut Edward Gibson tried to find out something about how a thin film of water would behave in microgravity. He put a drop of water on a wire loop. Then he slowly opened up the loop to see how thin a film he could make. What he saw was that the bubble became a mosaic of tiny, curved droplets—each one a perfect lens. The idea of making lenses and other glass products in space began with this discovery.

There will be many other discoveries. What will the future of space manufacturing be like? Let's jump ahead to the year 2025 and see.

5 | SPACE FACTORIES OF THE FUTURE

It is the year 2025. Several hundred space factories orbit earth, within a few hundred kilometers of the planet. Spacecraft fly back and forth between earth and the factories. Near earth, space is crowded, and traffic rules are strictly enforced—just as flight rules between major airports are.

Raw materials for the space factories making drugs and medicines come from earth. But for other space factories—those making high-quality glass products, computer parts, and alloys—that is not true. Most of these factories were not made on earth. Their raw materials do not come from earth, nor do spare parts for repairs.

All these things come from the moon. The moon has been important in making manufacturing in space successful.

The moon is over a thousand times farther away from the space factories than is earth. Yet it is far cheaper to send raw materials to the factories from the moon than to fly them up from earth.

The earth's gravity is about six times stronger than the moon's. That's why it took the enormous three-stage Saturn rocket to start the Apollo astronauts on their way from the earth to the moon. The rocket was 111 meters tall when launched from earth. Most of that length was simply storage space for the huge amount of fuel needed to get the rocket off the ground and into orbit. By contrast, the relatively feeble rockets on the lunar module were strong enough to lift the astronauts off the moon.

Another reason it's so costly to launch a spacecraft from earth is our atmosphere. Rockets lift off from earth almost straight up, to get through the thickest part of the air as fast as possible. The air acts as a brake, reducing the rocket's speed. The longer the rocket remains in the atmosphere, the more fuel is wasted.

On the airless moon, that is not a problem. Nowadays, in 2025, spacecraft headed for the near-earth space factories leave the moon without burning an ounce of rocket fuel. They glide along a level "railway" on the moon's surface. Electromagnetic forces hold the spacecraft suspended a short distance above the rails, so there is no friction. And the same forces push the rocket along at ever increasing speed. At last, the rocket catapults off the end of the railway at about 8,400 kilometers an hour and soars into space. It is free of the moon's gravity.

About 38,600 kilometers from the moon—a tenth of the way to earth—the earth's gravity takes over. The spacecraft is now going "downhill"—falling toward the earth. Earth's gravity is doing all the work. The spacecraft's rockets burn only for small maneuvers and to slow it down as it nears the orbiting factories.

In the 1980s, it cost about a thousand dollars to get a kilogram of material from earth's surface into earth orbit. The "moon catapult" could have delivered the same material from the moon to earth orbit for ten cents.

Space scientists of that day knew this. Plans for a moon catapult had been around for twenty years or so. But there were no permanent bases on the moon in the 1980s. No one had walked or driven on the moon for over ten years. The lunar buggy of the Apollo 15 moon landing in 1971 sat on the dusty surface like a junked auto.

The early moon walkers and moon riders had brought more than 360 kilograms of lunar soil to earth by 1972. That soil showed that the moon is rich in silicon and metals such as iron, aluminum, and magnesium. In fact, the moon has more of these materials on its surface than does the earth.

The moon, airless though it is, has plenty of oxygen. The oxygen is combined with silicon in the form of silicon dioxide. Silicon dioxide is common both on the earth and the moon as a part of many rocks and minerals such as quartz. Ordinary beach sand is nothing but very fine grains of quartz.

Silicon, as we've seen, is an important element in computer chips and other electronic devices. It is separated from the oxygen in the moon's soil by electricity from solar cells. The silicon is sent to space factories making electronic parts. The oxygen is used to supply air for the moon bases.

Sand itself is the basic ingredient of glass. Much of the

The lunar mass-driver ejects lunar rocks, gravel, and dust into space. These materials are then caught by a mass-catcher, and silicon, aluminum, and oxygen are removed from them. NASA

moon's sand is shipped directly to the space factories producing glass products such as optical fibers and lenses.

The moon now has factories of its own. Metals are mined and purified there. Spare parts for the orbiting space factories are largely made on the moon. The operation of the factories is computer controlled. Specific jobs are done by specialized worker robots. Unlike human workers, the robot factory worker is both worker and tool. Few factory robots look at all like people. Simpler robots of this type did much of the work at some automobile factories in the late twentieth century.

A few engineers and electronic troubleshooters live on the moon bases. They live on the side of the moon that faces earth, of course.

On the far side of the moon live the astronomers. There they can look deep into space without having the bright earth in the sky. And the moon shields their radio telescopes from the maze of radio signals that blanket the near side—messages to and from the earth, between spacecraft in transit, or between bases on the moon. The astronomers

A control tower for operations on the moon. NASA

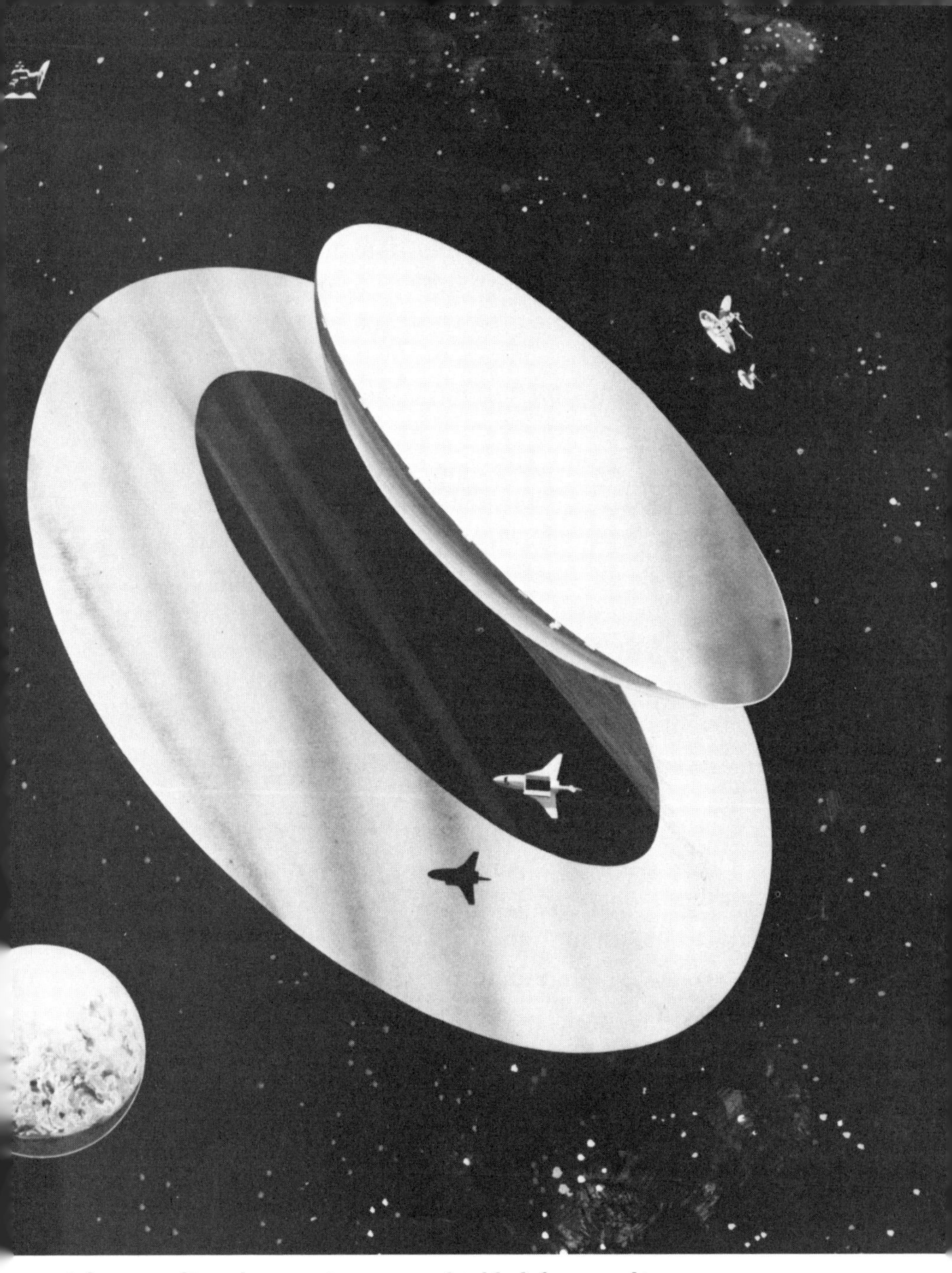

A large radio telescope in space, shielded from radio waves on earth by a disc-shaped device, will allow astronomers to pick up waves from outer space. NASA

on the far side have also put some distance between their telescopes and lunar pollution. The near side of the moon is beginning to have a very thin atmosphere of sorts. It is an atmosphere made up of waste gases from factory operations.

With factories on the moon and in space, pollution on earth is decreasing. Many factories have moved into orbit where heat and dangerous wastes can be gotten rid of without risks to people or to earth's environment.

In effect, the moon has become a planet-sized space station with its own natural resources. That, along with automation and reusable spacecraft, has made manufacturing in space a big success.

No single country on earth, however rich in money and in skilled people, can make manufacturing in space work on a large scale. That takes international planning.

The view from space helped encourage such cooperation. By the twenty-first century, many people had traveled in space. Men and women were coming into high office who had looked at our planet from space. They had seen earth whole, though they knew it was divided by hatred and fear, by greed and stupidity.

They had seen the stars undimmed by air, hard and bright against the blackness. They had walked in space, surrounded by stars and emptiness, and wondered: Where did it all begin? Do the answers lie somewhere out there? Space called to them.

But earth called, too. The small, lovely planet tugged at their hearts and minds with a pull as strong and as real as the gravity that held them in orbit.

They saw earth whole. Other space travelers, with television cameras for eyes and radio for voices, had brought the news back from the far reaches of the solar system. Of all the worlds that circled the sun, earth was the only likely home of intelligent life.

Because it is our home, it is worth preserving. Someday, we may travel far into space. But we will be wiser travelers for having learned how to preserve our home.

We've imagined what things will be like in 2025. Imagination can find many uses for space. Space is a new frontier, a place of adventure. In space, we may learn more about the universe and how it came to be. Space is a highway for worldwide communication and information sharing.

But space is also the "high ground" of the military—a place from which all nations are exposed to spying eyes and orbiting weapons. Fear of that exposure may bring about a different future in space.

6 | SPACE WARS

"In the next fifty years, hundreds of thousands of people will have gone into space and returned."

That's what Dr. Hans M. Mark told an audience at The Wheeler School in Providence, Rhode Island, in March 1982. Dr. Mark was at that time deputy administrator of NASA.

Many of the people Dr. Mark envisions traveling into space will make the round trip out and back aboard a space shuttle. But what will they do, while they are out in space? Will they be studying weather and climate patterns around the world? Keeping track of air pollution? Probing the far reaches of space with radio and optical telescopes? Improving methods of manufacturing in space?

Not if Dr. Mark and a number of other officials who plan the future of the shuttle have their way. They make no secret about it.

In that same speech, Dr. Mark pointed out that from its beginning the space shuttle has been "largely a military project." And in 1980, when Dr. Mark was secretary of the Air Force, he was even more blunt about the role of the United States space shuttle: "NASA is in fact a minor user and not the driver. That's something the NASA folks don't like to hear, but it's true." Verne Orr, the man who succeeded Dr. Mark as Air Force secretary summed it up: "Most people in the Air Force think far enough downstream we will probably be running the shuttle."

If that is true, there is no doubt what most space travelers—United States space travelers, anyway—will be doing. They will be on military duty.

The idea of war in space is nothing new. Any human invention can be turned to some warlike use. Rockets and space travel are no exceptions.

However, until 1957, space travel did not exist. Some scientists—and some science fiction writers—knew it would soon be a reality. But for most people, space travel was just science fiction.

On October 4 of that year, the exploration of space began. For the first time, a man-made object orbited earth. Once every ninety-six minutes, the artificial moon circled the globe. Instruments aboard the 83-kilogram unmanned spacecraft measured temperature and pressure. A radio sent the measurements back to earth.

The spacecraft was named Sputnik 1. *Sputnik* is Russian for "fellow voyager" or "traveling companion." For twenty-one days, Sputnik's beeping radio could be heard around the world. The satellite itself stayed up for three months.

The fact that the first spacecraft in history was made and launched by the Soviet Union caused panic in the United States. People asked: How did it happen? How had the "backward" Soviet Union beaten the United States into space? What was wrong with United States science and engineering?

The roar of Sputnik's rockets was the signal for the start of the so-called *space race,* a race that continues. On January 31, 1958, Explorer 1, the first United States satellite, was launched. In October of that same year, by an act of Congress, NASA was created.

The National Aeronautics and Space Act of 1958 says not one word about military activities in space. It made NASA a civilian space agency. Many of NASA's first astronauts were military pilots, but that was because they already had a large part of the flight and navigation training important for space travel. Also, they were in top physical condition.

These people did not orbit earth or land on the moon to gain a military advantage. It was an adventure, a challenge. It was a way to learn more about our own planet, our moon, and the solar system. It was a peaceful rivalry with Soviet cosmonauts and Soviet space probes.

We have learned a lot through that rivalry. People have explored parts of the moon. Unmanned spacecraft have mapped all of the moon, including the side not visible from earth. Probes have landed on Venus and Mars. Other remotely controlled probes have taken close-up photos of these two planets, as well as of Mercury, Jupiter, and its four large moons, Saturn and its ring system, and Uranus.

We've also learned much more about the world we live on—its weather, its oceans, and its lands.

Most of this new knowledge was shared among all nations. Pooling of information from weather satellites has made air travel safer and given advance warning of dangerous storms, for example.

Sharing information gained in exploring space is part of

NASA's job. The National Space Act that created NASA directs that NASA share as widely as possible information about its flights and the knowledge gained through them.

In fact, NASA did that for twenty-four years—from 1958 to 1982. Then there was a change. In the summer of 1982, the space shuttle Columbia made its fourth, and last, test flight. One purpose of the flight was a last checkout of the shuttle's ability to maneuver in space and on its gliding return to earth. Columbia also carried some experiments, including several related to manufacturing in space. But the biggest experiment aboard was booked by the United States Department of Defense. It was called simply *DOD 82–1*.

The name speaks for itself. *DOD* stands for the Department of Defense; *82–1* means the first military payload carried by the shuttle in 1982.

There will be many more. Through 1994, almost half of the scheduled flights—114 out of 234—will carry military payloads. The tenth flight, set for November 1983, will carry *only* military experiments. So will many flights thereafter.

As we've seen, NASA was created so the United States could develop a civilian space program. Yet at the same time, United States military experts were worrying about Sputnik, too. To these experts, space seemed a good place to park nuclear bombs. Such weapons, orbiting earth, could be commanded by remote control to reenter the atmosphere at a point where they would strike a chosen target.

The experts pointed out that a weapon launched from space would hit its target twice as fast as a missile fired from the ground. They were convinced that the Soviet Union would develop such bombs, as well as spy satellites and *antisatellites* (satellites designed to destroy "un-friendly" satellites).

Soviet progress in space seemed to give weight to the experts' fears. By 1961, the Soviet Union had sent two manned spacecraft into orbit. Vostok 1, launched on April 12, was piloted by Major Yuri Gagarin. Major Gagarin orbited earth once and then returned safely from space. On August 6, Vostok 2 went up, piloted by Major Gherman Titov. Titov made seventeen orbits of earth before landing safely.

Later, United States Air Force officers testified before Congress about the flight of Vostok 2: ". . . Major Titov's Vostok spaceship passed within one hundred fifty miles of this building," the officers told the congressmen.

Nevertheless, nuclear weapons were not put into orbit, either by the United States or by the Soviet Union. In fact, both nations signed a formal treaty promising not to place weapons of mass destruction in space.

Why did this happen? Mainly because both sides soon discovered that orbiting nuclear bombs were not going to be useful weapons. And they would create a lot of problems.

A bomb in orbit can only hit a target on its flight path. Changing orbits looks easy on the computer screen of a space wars game. But in fact it's a time-consuming process. It might take hours or days to hit a given target from space, once the decision has been made. And then the bomb might miss by many kilometers. Remote control of such weapons from the ground is unreliable.

Worse still, satellites may develop problems that make them hard to control. As a result, they may reenter the atmosphere out of control. That's what happened to the United States Skylab, which fell in bits and pieces over the

Australian desert. In the same way, the Soviet Cosmos 954 came apart over the Canadian Arctic. Radioactive parts from the spacecraft's reactor were scattered over the land.

This reactor was fueled by plutonium, a common power source in the nuclear reactors used on some satellites. Plutonium is one of the most deadly poisons known. It is highly radioactive. It tends to build up in bones when absorbed into the body. Plutonium's half-life is about 24,400 years. In other words, it takes that long for the metal's radioactivity to be cut in half.

This was not the first time a plutonium-fueled reactor broke up in the atmosphere. On April 21, 1964, a U.S. navigation satellite was launched from Vandenburg Air Force Base in California. On board was a reactor fueled by about one kilogram of plutonium. The rocket engines failed before the satellite reached orbit, and the satellite plunged earthward over the Indian Ocean.

But it never reached the ocean. Search teams concluded that the satellite had burned up completely in the atmosphere.

That did not alter the radioactivity of the plutonium. Radioactivity is not affected by the chemical changes involved in burning. The plutonium spread through the upper air as a fine dust. Most of this dust was concentrated south of the equator.

Like a cloud, the plutonium dust floated in the air and drifted with the winds. But clouds are made of water. They disappear when the water falls as rain or evaporates.

The invisible cloud of plutonium dust did neither. It slowly spread worldwide. Not until 1970 had most of it fallen out of the atmosphere. Scientists estimated that this one accident caused plutonium contamination around the world to triple.

But space does have some very practical military uses. Navigation and guidance satellites pinpoint the movements of planes and missiles. So-called *spy satellites* can help enforce nuclear disarmament agreements and prevent cheating. Such satellite systems can also give warning of a pending missile attack. This makes it harder for either side to launch a surprise attack.

These satellites are important both to the United States and to the Soviet Union. Both sides worry about their satellites being vulnerable to antisatellite weapons. At the same time, both sides are trying to develop their own antisatellite weapons, as well as to improve their defenses.

For the United States military, the shuttle has been the ideal tool for giving us the ability to attack and defend in space. The shuttle can "park" extra spy satellites in safe, high-altitude orbits where they can quickly replace damaged satellites. The shuttle crew itself can repair some damaged satellites in space. And in the future, the Department of Defense expects to use the shuttle to test laser beams and high-energy particle beams. These, DOD hopes, will be useful antisatellite weapons.

Many experts disagree. They say laser weapons would be costly and unreliable. Such weapons, and indeed all satellites and spacecraft, would be vulnerable to space mines.

A space mine is a high-powered nuclear bomb exploded in space. The explosive force of such a blast in space would not be very destructive. There is no air to carry the shock wave from the blast. But high-energy radiation—X rays and gamma rays—would spread out in all directions in space at the speed of light. On earth, this radiation is weakened and absorbed by the air.

This radiation will cause high voltages and large electric currents wherever it strikes metal. Vital computers and

electronic circuits on board satellites will burn out almost immediately when the radiation hits them.

DOD claims that tests done on the ground show that satellites can be protected against such mines. The actual results of these tests are secret. Some scientists who have seen these results do not agree that they prove satellites can be protected.

DOD still presses on with the testing of laser weapons. On March 23, 1983, President Ronald Reagan endorsed the search for antisatellite weapons in a nationally televised speech.

The shuttle, from its beginning, was designed with military uses in mind. It was to be more than a space bus and repair truck for military satellites. Eventually, if the tests work out, it will be the heart of a command post for a kind of laser battle station in space. DOD plans to use nuclear reactors to power these stations and other military satellites.

How did this happen? How did the shuttle become so deeply involved in the military development of space?

The idea for the space shuttle began in September 1969. Just three months before, the first men had walked on the moon. That was one of NASA's great successes.

Yet NASA and the civilian space program of the United States were already in lots of trouble. To many people, the space program seemed like a huge waste of money—money that could be better spent helping to solve our problems here on earth, problems such as poverty, pollution, malnutrition, crime, and so on.

Many of those who felt that space exploration was valuable and useful thought the manned flights to the moon wasteful. They believed much more could be done less expensively through unmanned flights. And, in fact, the un-

manned probes that explored the solar system during the 1970s proved that over and over.

During that time, NASA, by itself, could not get a manned space program such as the shuttle approved. But with the help of DOD, it could. While NASA's budget was going down, the space budget of DOD was going up. It has been going up every year since 1972. By 1982, DOD's *space* budget was bigger than NASA's!

In fact, if it weren't for the influence of DOD, the shuttle would probably never have been built. Just before the first shuttle flight in the spring of 1981, John Yardley, NASA's director of space transportation, said, "We did need the support of the Department of Defense and the Administration to get the money [for the shuttle]."

Not that DOD proposed giving any of its money to NASA for developing the shuttle. Instead, DOD used its influence with Congress and the president to get NASA's funds for the shuttle approved. DOD pointed out that the shuttle might have many military uses. In return, NASA agreed to redesign the shuttle to fit in with DOD's plans for space.

The original NASA shuttle design called for a small, stubby, oval-shaped shuttle, without wings. Instead of getting lift from wings, the whole spacecraft acted as a wing on its steep, fast glide down to earth.

But DOD wanted the shuttle to be able to glide at least 1,600 kilometers cross-country on its return from space. For that, the shuttle needed wings. And the cargo bay had to be made much bigger so that it could carry large military payloads.

Why does the shuttle need this gliding ability? The reason is that DOD wants to launch certain satellites from Vandenberg Air Force Base in California. The shuttle is to

pick up these satellites one orbit later, and return them to the California air base.

These satellites will be in polar orbits. A satellite in polar orbit passes over the North Pole and the South Pole on each trip around the world. A globe of the world will help you see what this means.

The globe is supported by a half circle of wood or metal. The half circle is attached to the globe at the North and South poles so that it can turn. Imagine the half circle extended to make a full circle around the globe. This is the orbit of a polar satellite.

As a polar satellite travels around the earth, the earth is turning beneath it from west to east. After one orbit, a satellite launched from California will pass over the Pacific Ocean about a thousand miles west of Vandenberg. This is where the shuttle will pick up the satellite and return it to Vandenberg. To do that it must be able to glide to the base after reentering the atmosphere. So the shuttle needed wings.

DOD does not say what the purpose of such satellites would be. But a map of the world shows that they would travel on a north-south line that passes over the western USSR, Iran, the Persian Gulf, and Saudi Arabia. The Persian Gulf is a vital seaway for most of the oil used by the U.S. and its allies.

For scientific purposes and for experiments in space manufacturing, the shuttle does not need to be nearly so large. The extra size and weight make shuttle flights more expensive. This makes it harder for the shuttle to compete with experiment-carrying rockets from other countries. We'll see in the last chapter that there is going to be plenty of competition.

Gradually, the Department of Defense's influence on the shuttle has become greater. Since the shuttle now has a large military role, DOD naturally wants to have more say about what flies on the shuttle at what time. So DOD reserves the right to "bump" (remove) civilian cargo from a shuttle flight at any time. NASA protested against this policy, but a Space Review Board, appointed by President Reagan, overruled NASA. All DOD has to do is to say it needs the space for reasons of national security, and space will be made.

Companies and governments pay many millions of dollars for shuttle space. If they are bumped, they are likely to decide the shuttle is not a good investment.

As DOD influence over the shuttle increases, the shuttle is less able to help the United States compete with foreign development of manufacturing in space. Even now, the Spacelabs to be used by the shuttle are being built in Europe because NASA can't spare the money to build them.

But it is a vicious circle. NASA is trying to offer the shuttle as a carrier for the European-built Spacelabs. And yet, at the same time, DOD's interference with shuttle scheduling is making European and other foreign investors more and more doubtful about the shuttle's usefulness.

7 WHICH FUTURES?

The last test flight of the space shuttle lifted off in the summer of 1982. It was a historic flight for many reasons.

First, it showed that the shuttle is both a reliable spacecraft and a reliable aircraft. Astronauts Thomas K. Mattingly and Henry W. Hartsfield flew the space shuttle Columbia around the earth 113 times—a distance of 45,052,000 kilometers. Then they brought it down to a flawless landing on July 4, 1982.

Secondly, the shuttle carried a secret military payload into space. That had never been done before by the United States.

Lastly, Columbia also carried a package of electrophoresis experiments, designed by scientists at Johnson & Johnson.

The results of the experiments delighted the scientists.

United States

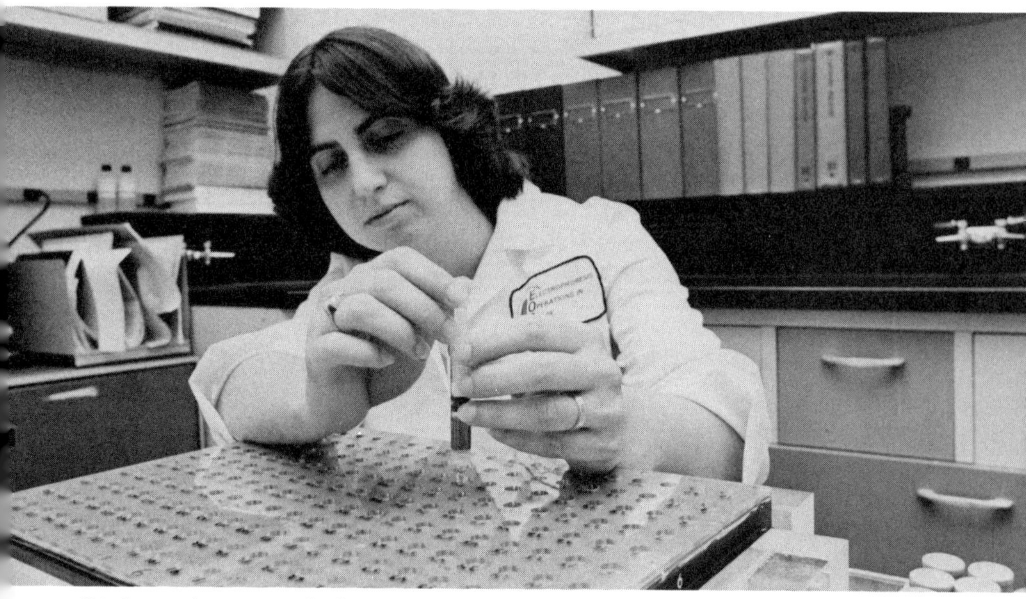

Biological material that was separated on the space shuttle's fourth test flight is analyzed by a McDonnell Douglas technician. The tests showed that electrophoresis can process about five hundred times more material in space than on earth. MC DONNELL DOUGLAS CORPORATION

Microgravity electrophoresis worked far better than earth-based electrophoresis. In space, four hundred times as much medicine was produced, and the product was five times purer than the same medicine made on earth.

That may be the most important thing about the flight. The electrophoresis experiments showed that commercial manufacturing of biologicals in space would be possible within a few years.

That's good news. Yet the very success of these experiments may make it harder for the United States to make a success of manufacturing in space.

◀ *The space shuttle Columbia.* NASA

Let's see why this is so.

July 4, 1982, was bright and clear at Edwards Air Force Base in the Mojave Desert of California. Among the thousands of spectators straining for a first glimpse of the Columbia was President Reagan.

An hour earlier, Columbia had been a spacecraft in orbit. Halfway around the world, in darkness, 275 kilometers above the Indian Ocean, Columbia flew backward. For two and a half minutes her engines fired, slowing her down. Columbia began to fall out of orbit on her long path home.

Half an hour later, 129 kilometers over the western Pacific, Columbia became an aircraft. She reentered the atmosphere. The thickening air gripped the Columbia like a brake slamming on. And brakes make heat.

Within minutes of reentry, the nose and leading edge of the wings heated up to 1500 degrees Celsius (2700 degrees Fahrenheit). Columbia was surrounded by a glowing fireball of air. Mattingly and Hartsfield put their helmet sunshades down to shield their eyes from the glare.

Over the Pacific Ocean, the real flying began. Gliding through the air, traveling at over 28,900 kilometers an hour, Columbia made a series of S-shaped turns, north and south of the straight-line path to the landing strip. The winding path helped slow the spacecraft-glider down to landing speed.

To those on the ground at the Air Force base, Columbia was heard before it was seen. At 3,100 kilometers an hour—two and a half times the speed of sound—Columbia caused a thunderous sonic boom as it hurtled over the California coast. Seconds later, a speck appeared in the sky. It grew rapidly in size as Columbia approached the runway in a steep, silent glide. Columbia leveled off at 534 meters. Now the shuttle was fully visible, an aircraft as big as a me-

dium-sized jetliner, but gliding to a landing without a sound except the rush of air past its wings.

The president made a short welcoming speech to the astronauts. "The fourth landing of the Columbia is the historical equivalent to the driving of the golden spike which completed the first transcontinental railroad," he said. ". . . The test flights are over, the groundwork has been laid. Now we will move forward to capitalize on the tremendous potential offered by the ultimate frontier of space."

It was time, the president went on, to turn the work of manufacturing in space over to private business. NASA had created a space transportation system. "Columbia and her three sister ships will be . . . ready to provide economical and routine access to space for . . . commercial ventures." And business will leap into space.

Or will it?

To planners at NASA, talk about the golden spike and the building of the great transcontinental railroads has a familiar ring. They themselves have compared the building and testing of the shuttle fleet to the building of the great railroads.

In an earlier time, those railroads helped the United States to become a great industrial nation. The railroads brought the raw materials to the factories. They brought the finished products to the buyers. And what the railroads did for ground-based factories, the shuttle can do for space factories.

The railroads were built by private companies. But these companies had a lot of help from the federal government. Among other things, the government owned much of the land on which the railroad tracks were laid. And it sold that land to the companies very cheaply.

In the same way, NASA officials feel that the United

States government must encourage private business to put money into space factories. Many business leaders agree. It can pay for more research into manufacturing in space. Such research will uncover and solve some of the unexpected problems that turn up in any new field. That will make it less risky to invest money in space factories. Or the government can lower taxes on businesses that go into space manufacturing.

Help like this is needed for at least two reasons. For one thing, space factories are expensive, and it will be years before they begin making money. Just as important, United States businesses face strong competition in space. European and Japanese companies are very interested in manufacturing in space. And these companies are getting a lot of help from their governments.

Since 1972, Europe has had its own space agency, the European Space Agency (ESA). By 1982, ESA had thirteen members. Belgium, Britain, Denmark, France, West Germany, Italy, the Netherlands, Spain, Sweden, and Switzerland are full members of ESA. Austria, Ireland, and Norway are associate members.

Already, ESA is competing with the United States in space. ESA offers its services with ads like this in a 1981 issue of an American publication, *Aviation Week and Space Technology*:

ARIANESPACE
FIRST COMMERCIAL OPERATIONS
SPACE CARRIER

Ariane is ESA's answer to the shuttle. Unlike the shuttle, Ariane is an "old-fashioned" three-stage rocket. It is not reusable.

ESA's experiments with Ariane were encouraging until mid 1982. Then, a launch failed. European space officials believe a guidance system was at fault.

Yet Ariane may still succeed. If it does, it should have one big advantage over the shuttle. It will probably be less expensive to use. ESA gets so much support from European governments that Ariane should be able to charge relatively low rates.

Equally important, Ariane's schedules are likely to be more certain than the shuttle's. And Ariane will not be bumping civilian cargo in favor of military payloads.

Another ESA project is Spacelab. Spacelab began as a joint undertaking by ESA and NASA. ESA designed and built the Spacelabs. NASA gets one complete Spacelab free. In return, ESA gets to use the shuttle free of charge for its first Spacelab flight, scheduled for 1983.

The countries of Western Europe had to cooperate to design and build Spacelab. The Spacelab program made ESA a stronger organization. It united Western Europeans in a common effort to explore and use the resources of space.

So, in a sense, the NASA-ESA Spacelab deal taught Western Europe the value of cooperation in exploiting space. As a result, ESA and NASA may one day become rivals in space.

Or they may not. European space officials know there are good reasons to continue cooperating with the United States space program. One reason is that the United States has an edge over other nations in rocket and spacecraft design. The shuttle, in spite of its problems, still has many advantages over expendable rockets. The shuttle's speed increases more slowly at liftoff. That puts less strain on delicate parts of satellites and instruments in the cargo bay. So they don't need bulky expensive protection. Astronauts can check out problems with cargo after the shuttle is in space. That means there does not need to be a lot of remote-control equipment carried aloft.

However, the shuttle cannot make the most of these advantages. Because of budget cuts and space reserved for military cargo, the shuttle can't give maximum transportation to private companies. By 1982, the shuttle schedule through 1985 was cut from forty-eight flights to thirty-two flights.

Other international efforts are also possible in space. ESA, Canada, and Japan are interested in building a permanent space station with the United States. Such a station, as we saw in chapter four, would give a big boost to manufacturing in space. It would encourage private companies to invest in space factories, some scientists say.

The United States, Japan, Canada, and the nations of Western Europe are not the only countries with an interest in the peaceful uses of space. In the spring of 1982, the Soviet Union launched the Salyut 7 space station. It replaced Salyut 6, which had been in use since September 29, 1977. Salyut 6 was resupplied regularly by an unmanned, remote-controlled spacecraft. The two spacecraft docked while the crew of Salyut 6 brought new supplies aboard. That allowed the crew of Salyut 6 to stay in space for months at a time. In 1980, a crew stayed aboard for 185 days, a world record. During that time, the crew performed many hundreds of experiments in space manufacturing. Two years later, another Soviet astronaut team stayed in orbit even longer.

Western scientists believe that Salyut 7 is a larger version of the Salyut 6. They say it may be the next step toward putting a permanent Soviet station in space. As evidence, they point to the words of Soviet scientist Aleksey Yeliseyev, who said in 1981, "We are ready to make the next step—to graduate to the creation of permanent orbital complexes [space stations]."

By the mid 1980s, a permanent Soviet station may be in orbit. Such a station will have some military uses. But United States experts are sure it will also be used for experiments in manufacturing in space. United States officials say that about three hundred fifty Soviet scientists are already working on the design of those experiments.

Some people in the United States space program hope that the Soviet space station is launched soon. They feel it will give a big push to United States space efforts—just as the launching of the first Sputnik did back in 1957.

That is possible. But which way will it push the United States space program? Into developing space for manufacturing and other peaceful uses? Toward building up a strong, healthy United States industry in space?

Or will the United States put the bulk of its money, time, and effort into building up military bases in space? That would leave the industrialization of space to the Europeans, the Japanese—and quite possibly to the Soviets, too.

There is still another possibility—the peaceful use of space by private companies. One such company is Spacetran.

Spacetran, based in Princeton, New Jersey, wants to buy a shuttle from NASA. The price would be about a billion dollars. Spacetran's president is Klaus Heiss. Heiss is an economist who worked for NASA in the 1970s. He helped NASA analyze the costs of designing and building the shuttle.

Dr. Heiss predicts that almost half of NASA's shuttle flights will be taken up by the Department of Defense. Other United States government agencies—weather forecasting services, for example—will also be using the shuttle. According to Dr. Heiss, that won't leave enough room aboard for companies that want to make use of space for commercial purposes. That includes, of course, manufacturing in space.

His idea: Use Spacetran to carry the commercial customers.

A group of companies, including insurance agencies and banks, like Dr. Heiss's idea. They are ready to lend him the money to buy the shuttle from NASA. If the government okays the deal, Heiss expects to have a working commercial shuttle by 1987. He plans to hire NASA to launch his flights. United States companies might pay Heiss for space aboard the shuttle.

And Heiss also expects to get foreign customers for his shuttle. He won't have to worry about budget cuts or the Department of Defense. So the Spacetran shuttle should be able to keep to its flight schedules—and still have the advantages of the shuttle. A private shuttle might well steal some customers from Europe's Arianespace.

Spacetran could make President Reagan's dream about the future of United States industry in space come true: Private companies pay their own way into space. They compete to build the best factories and earn the most rewards. Everybody wins. The companies that do well earn the most. Their products get cheaper and better. Their factories provide more jobs. Building and improving the factories provide still more jobs.

Will this happen?

Can Spacetran—or another private American space company—meet this kind of competition without support from the United States government? Only the future will tell. Space factories have many possible futures. Some will come true.

Factories in space will be part of *your* future.

INDEX

Page numbers in *italics* refer to captions.

ABOUT THE AUTHOR

Author MALCOLM E. WEISS says: "Experiments in space manufacturing show that there may soon be working factories in space. On a small scale, space manufacturing has already begun. Exciting possibilities lie ahead. How will space factories be built? Who will build them? We have the imagination and the tools to make new materials and medicines. How and when we do it depends mainly on whether nations compete or work together in orbit—a place where oceans and national boundaries are crossed every few minutes."

This is the author's sixteenth book for young people. He and his wife, also a writer of children's books, live with their two daughters in Maine.